I0477251

Biology serves as the study of life and focuses on understanding how life exists, develops, and changes. Biology has many useful applications including improvements to food production, treating diseases and other medical conditions, and conservation of species to name a few.

There are eight characteristics that are common to all forms of life. They are as follows:

1. All living things are composed of cells. These cells are the simplest forms of life.

2. All living things use energy in some form to perform tasks.

3. All living things interact and are acted upon by their

environment.

4. All living things maintain a state of homeostasis which means that the living thing is a separate entity from its environment.

5. Living things grow and develop over time.

6. Living things have genetic material which is used for the reproduction of living things.

7. Organisms of a population evolve from one generation to the next passing on genetic information during reproduction.

8. All living things are connected by a complex evolutionary history.

There are a few other important concepts that can be seen throughout biology in addition to eight main characteristics. For instance, Structure determines the function of organisms and parts of an organism. Also, there are new properties of life that emerge time through complex interactions such as evolutionary changes and dynamic interactions between organisms and their environment. Lastly, biology has impacts on other areas of that are study such as social behaviors, individual qualities, and societal development.

When studying biology, one can examine biological organisms at different levels. The levels of biological organization include atoms, molecules, macromolecules, tissues, organs, organisms, populations, communities, ecosystems, and the biosphere. The different levels of biological organization are not separate from or act independent of the other levels. These levels operate systemically with changes and properties of lower levels affecting the events that occur at higher levels of organization and vice versa. This type of inter-play is dynamic in nature, design, and function.

The changes that occur in species are the result of the modifications of structures that existed previously. This can often be noted as commonly seen in vertical evolution when a mutation creates a specific characteristic that is beneficial to the survival of species. The mutated expression of the species then has a higher likelihood of producing offspring. The offspring will have the genetic information that was passed along from the parent that will increase its likelihood of producing more offspring. The mutated variants of the species will increase in numbers eventually become common enough through breeding with the non-mutated members of the species to where the trait will become a common characteristic throughout the entire population over time. This process is called natural selection. In addition to vertical evolution, horizontal gene transfer has been noted at a cellular level between different species. These process interact to create the complexity of life throughout the ecosystems and biosphere.

Taxonomy is defined as the grouping of species based upon their evolutionary relations to each other. Taxonomy narrows species from broad groups starting with the domain and specifying the individuals species narrowing groups down to the specific species. Domain, which consists of three groups, is comprised of the domains Bacteria, Archaea, and Eukarya. The organization of groups from most general to most specific is as follows:

1. Domain

2. Kingdom

3. Phylum

4. Class

5. Order

6. Family

7. Genus

8. Species

The diversity in species has developed over millions of years through mutations in the genetic codes of living organisms during reproduction. The genetic composition of a species is referred to as the genome. The genome is the genetic blueprint that is passed from the parent to the offspring. The genome carries the information that is necessary for the expression of all the various traits of living organisms. The mutations of such a blueprint is the key process for evolutionary change over generations. Through the genome, one can note that there is a pattern that is essentially the design for the way in which proteins are created by a cell. The collection of proteins produced by a cell is called to the proteome.

Foundational Chemistry Related to Life

All material can be divided down into increasingly smaller units of matter until one reaches the level of the atom. The atom is smallest unit of matter that cannot be further broken down without losing its chemical properties. Atoms are composed of protons, neutrons, and electrons. Hydrogen atoms are the only atoms that do not have neutrons, but they do have one proton in their nuclei. The electrons orbit the nucleus of the atom. Protons have a positive charge, neutrons have a neutral charge, and electrons have a negative charge.

Each element contains a specific number of protons that relate to its atomic number displayed on the periodic table of elements. The periodic table serves as a clear guide on the organization of elements based on their atomic numbers and designed based on the number of electrons in their orbital shells. In addition to the number of protons , the periodic table displays the average mass of each element as measured in atomic mass units. Atoms in nature commonly exist as isotopes. Isotopes contain differing numbers of neutrons. Some of the isotopes become unstable due to the number of neutrons that they have in their nuclei. This instability causes them to emit radiation for the purpose of returning to a stable isotopic state. Most living organisms are primarily composed of oxygen, carbon, hydrogen, and nitrogen atoms. There are numerous other elements that are present in living organisms in trace amounts that are essential for growth, development, and function.

Molecules are two or more atoms bonded together. When bonded together, these molecules have different properties than the individual elements that they are composed. A compound is composed of two or more different elements' atoms. Atoms have a tendency to form bond with other atoms using the electrons in their outer shells to fill the outer shells of the atoms. Covalent bonds form when atoms share the electrons of their outer shell. Covalent bonds are strong bonds. When two electrons are shared between two atoms, the bond is called a double bond; when one is shared, it is a single bond; and when three are shared, it is called a triple bond.

The measure of an atom's ability to attract bonded electrons is called electronegativity. Atoms form polar covalent bonds when the atoms that the distributions of the electrons across the atoms are more attracted to one side of the molecule than the other. The bond that hold water together are examples of a polar bonds. Polar bonds also create a difference in charges across the molecule. It is important to be aware that polar molecules attract other polar molecules and repel non-polar molecules.

The association of the interactions between polar molecules create van der Walls forces which are weak electrical attractions that arise between molecules as the orbit one another. The oppositely charged ends of the molecules attract each other, and the molecules tend to stay together in groups. A specific example of this attraction is the attraction between the positively charged atoms in one compound of water to the negatively charged oxygen atoms in other compounds of water. This creates the tension on the surface of water and several other liquids. The weak bonds as a result of the positively charged hydrogen to the negatively charged atoms of other molecules are called hydrogen bonds. Another type of commonly occurring bond between atoms is an ionic bond. Ionic bonds occur when one atom gives up an electron to another atom. The atom that gives up the electron is called a cation and has a positive charge. The atom that gains the electron is call an anion and has a positive charge. The both atoms are called ions and have some type of electrical charge. Molecules that are unstable and interact with other molecules by taking away electrons from their atoms are called free radicals. Chemical reactions are the interactions of different substances that occur when atoms of compounds are exchanges, separated, or combined. Chemical reactions achieve an equilibrium in which the reacting compounds are equal to the products of the reaction.

In most living organisms, water serves as the solvent for most chemical reactions that are occur both inner-cellular and extra-cellular. When atoms are dissolved in water, they interact in ways that would not be possible without the atoms being dissolved in the water. The molecules that dissolve in water are called solutes. Once the molecules are dissolved the mixture is called a solution. The solute concentration is the amount of the solute that is dissolved in the solvent. Solution molarity is the number of moles dissolved in one liter of the solution. Molecules that are ionic or polar tend to be hydropilic (water-loving). Hydrophillic molecules dissolve in water. Non-covalent molecules tend to be hydrophobic (water-fearing) and do not dissolve in water.

Water serves as a very stable liquid platform for reactions due to its high heat of vaporization and high heat of fusion. Water molecules also participate in many of the chemical reactions that occur in living organism by exchanging hydrogen and oxygen atoms with other molecules. One common chemical reaction that occurs involving water is hydrolysis. In hydrolysis, large molecules are broken down into smaller molecules through there interaction with water. Dehydration reactions are the opposite of hydrolysis. In these reactions, larger molecules are broken down into smaller molecules. Water is also used in living organisms to perform various essential functions to live such as dissipating body heat, waste elimination, the movement of liquid through vessels, lubrication, and its surface tension can also serve as a surface for some insects to walk on.

The pH scale measures the concentration of hydrogen ions. Substances with a pH below 7.0 are acidic, and those with a pH above 7.0 are called alkaline. Pure water has a pH of 7.0 allowing it to serve as a neutral substance in relation to pH. With regards to the various properties of water, it tends to serve a role of high importance in living organisms' chemical reactions.

Organic Molecules

Organic molecules are molecules that contain carbon. Many common organic molecules such as lipids and other large complex molecules are called macromolecules. The study of molecules containing carbon is called organic chemistry. Carbon can form up to four covalent bonds with other atoms. When atoms consist mostly of hydrogen-carbon bonds, the molecule is called a hydrocarbon. Hydrocarbons are hydrophobic and poorly dissolve in water. Most organic molecules contain functional groups. Functional groups are groups of atoms with characteristic chemical properties and features. Here are some common functional groups:

Functional Group	Formula	Found in	Properties
Amino	NH_2	Amino Acids	Weakly basic, polar, forms part of peptide bonds
Carbonyl-Ketone	CO	Proteins, steroids, waxes	Polar, very reactive, forms hydrogen bonds
Aldehdye	CO	Odor molecules and linear sugars	Polar, highly reactive, forms hydrogen bonds
Carboxyl	$COOH$	Fatty acids and amino acids	Acidic, forms part of peptide bonds
Hydroxyl	CH_3	Amino acids, carbohydrates, alcohol, steroids	Polar, forms hydrogen bonds with water
Methyl	OH	Attached to DNA, carbohydrates, and proteins	non-polar
Phosphate	PO_4	Nucleic acids, phospholipids, and ATP	Polar, weak acid, egatively charged in living organisms

Sulfate	SO_4	Attached to lipids, proteins, and carbohydrates	Polar, negatively charged
Sulfhydryl	SH	Proteins with cysteine	Polar, forms disulfide bridges in proteins

Structural isomers contain the same atoms that are bonded in different ways from each other. Stereoisomers have the same bonding relationships but position their atoms differently. The two types of isomers are cis-trans isomers and enantiomers. Enantiomers are pairs of molecules that are mirror images of each other. For example, there are four different ways that an atom can bond to a carbon atom. In the case of the molecules being mirror images, they would be enantiomers. Cis trans isomers involve rotating two atoms that are bonded a carbon-carbon chain at differing angles. For example, in one isomer, the atoms may both be on the same side of the chain, but a different isomer may have the two atoms on opposite sides of the carbon-carbon chain.

Large macromolecules are created by linking together smaller molecules. The smaller molecules are called monomers, and the larger molecules are called polymers. The process in which two or more molecules combine to form a larger molecules is called a condensation reaction. In organic molecules, this process is called a dehydration reaction because it typically involves the release of a water water molecule when the monomers combine. Hydrolysis separates the monomers and absorbs a water molecule as mentioned before.

Carbohydrates are composed of carbon, hydrogen and oxygen atoms. Sugars are small carbohydrate molecules. The simplest sugars are monomers that are called monosacchrides. Sugars that have five carbon atoms are called pentoses. Pentoses are ribose and closely related deoxyribose which makes up the "D" in DNA. When two monosaccharides are combined, they form a disaccharide. The bond between the two sugar monomers by the dehydration reaction is called a glycosidic bond. When many monosaccharides are joined together, they form polysaccharides such starches and glycogen. Some notable polysaccharides include cellulose, chitin, and glycosaminoglycans which is found in cartilage and bone tissues of animals.

Lipids are hydrophobic molecules that are non-polar and account for 40% of the organic matter in the average human body. Triglycerides, which are formed by lipids, are commonly referred to as fats. These fats are linked by a chain of carbon and hydrogen atoms with a carboxyl group at one end. When all the carbons in the fatty acids have single bonds, the fatty acid is called a saturated fatty acid. In some fatty acids, a few of the carbon-carbon bonds are double bonds. These fatty acids are called unsaturated fatty acids. Phospholipids are similar to the triglycerides except that a phospate group is linked to the end of the carbon chain. Phospholipids also tend to have a positively charged nitrogen based compound attached to the phosphate group. This causes the phospholipid to be a polar molecule and hydrophillic. Steroids are another type of lipid but is much different from the other lipids. Steroids have four fused rings of carbon that form the skeletal structure of steroids with a hydroxyl group at the end of the ring structure.

Proteins are polymers found in cells that play key roles in almost all process of life. Proteins are made up of carbon, hydrogen, oxygen, nitrogen, and a few other elements in small quantities. Proteins are composed of amino acids . Amino acids are compounds with a carbon atom that is linked to an amino group, an amino acid side chain, a hydrogen atom, and a carboxyl group. Amino acids are joined together by dehydration reactions. In these reactions, carboxyl group of one amino acid is joined to the amino group of another amino acid. The bond between the carboxyl group and amino group is called a peptide bond. When multiple amino acids are connected by these bonds, a polypeptide is formed. When long chains of amino acids form, the end with the free amino group is called the amino end, or N-terminus. The end with the free carboxyl group is called the carboxyl end, or C-terminus.

Proteins have four levels of structure that are progressively more complex. The first level, or primary structure is the amino acid sequence. The secondary structure is the repeating patterns of folds of the amino acids with in the protein's structure. The tertiary structure is the folding of the protein structure itself into a complex three-dimensional shape overall. The final structural level, the quaternary structure, refers to organization of multiple polypeptide chains that combine to form the protein overall. As one notes the complexity of the levels of proteins structure and organization, a high level of complex design can be noticed. The individual polypeptide chains are called protein subunits. The proteins formed by multiple polypeptide chains are called multimeric proteins. Hemoglobin is an example of multimeric protein.

Nucleic acids are the essential organic molecules responsible for the transfer of genetic information. There are two types of nucleic acids, deoxyribonucleic acid (DNA) and ribonucleic acid (RNA). DNA molecules store genetic code, and RNA molecules serve to decode the genetic material for cell function. Nucleic acids are polymers that are composed of monomers arranged in a repeating sequence. The monomers of nucleic acids are called nucleotides. Each nucleotide has three components which are a phosphate group, a pentose sugar, and a single or a double ring of carbon and nitrogen atoms that are referred to as a base. DNA has four different nucleotides. They are adenine (A), guanine (G), cytosine (C), and thymine (T). Adenine and guanine have a double rings of carbon and nitrogen and are called purine bases. Cytosine and thymine have a single-ring structure and are called pyrimidines. RNA uses ribose in each nucleotide as opposed to the deoxyribose that is used in DNA. Additionally, the thymine is replaced with uracil. The similarities in RNA and DNA allow for RNA to use the sequence of the DNA nucleotides to mimic the patter for the purpose of protein replications based the pattern present in the DNA molecules.

Cell Features

Cell biology is the study of individual cells and the interactions of cells with one another. Cell theory explains that all living organisms are composed of one or more cells,

cells are the smallest units of life, and new cells only come from pre-existing cells.

The microscope is a magnification tool that allows researchers to study cell structure and function. An image taken with a microscope is a micrograph. A compound microscope is one that has more than one lens. The first compound microscope was invented in Holland in 1595 by Zacharias Jansen. There are three parameters of microscopy. Resolution is the measure of the clarity of an image. Contrast is the ability to visualize a particular cell structure depending on how it looks different from surrounding structures. Magnification is the ratio between actual size and the size of the image produced by a microscope. Light microscopes utilize light for illuminating an image, and electron microscopes utilize a beam of electrons. There are two types of electron microscopes, transmission electron microscopy (TEM) and scanning electron microscopy (SEM). TEM uses a beam of electrons transmitted through a biological sample that is stained with a heavy metal to provide a cross-sectional view of the sample. SEM uses a beam of electrons to view the surface of a sample that is coated with a thin layer of heavy metals.

The internal operations of a cell are the result of the cells ability to build and maintain such an organization through the use of proteins that bind to each other to create cell structures and facilitate processes. These type of interactions are called protein-protein interactions. The genome is the code that every species has that serves as the blue print for the creation of such proteins for these interactions. The genome of each species is comprised of genes that carry the information for the production of each type of protein that the cells need to produce.

There are two categories of cells, which are eukaryotic cells and prokaryotic cells. Prokaryotic cells are found in bacteria and archaea. Prokaryotic cells have a kernal-like appearance. They have a plasma membrane which serves as a barrier between the cell and its environment. This membrane is composed of a phospholipid double layer and embedded proteins. The inside of the cell is called the cytoplasm. The structures of the bacteria are contained in the cytoplasm. Inside the cell, the nucleoid region is where the DNA is stored, and ribosomes are the structures that are responsible for the production of polypeptides.

Bacteria have a cell wall surrounding the cell to provide it with structural support and protect the plasma membrane and cytoplasm from the things in its environment. To help bacteria from drying out, they secrete glycocalyx which is an outer viscous covering surrounding the bacteria that traps water. Certain bacteria develop a think, gelatinous glycocalyx layer called a capsule that help them avoid being destroyed by the immune system of animals when the infect host animals. This capsule also helps them to attach to cell surfaces when coupled with the pili which are appendages that project from the cell. In addition to the pili, bacterial cells also have another type of appendage called flagella that provide the bacterial cell with a way to move around their environment.

All species, with the exception of bacteria and archaea, are eukaryotes. These include plants, animals, fungi, and protists. Eukaryotes distinguish themselves from prokaryotes by having membrane bound organelles that perform specialized functions. An organelle is a membrane-bound compartment in each cell with its own unique function and structure. The separation of different regions that exist in eukaryotic cells is called compartmentalization. Most of the same types of organelles are found all eukaryotic cells. These organelles include the nucleus, golgi apparatus, endoplasmic reticulum, and mitochondria to name a few.

Eukaryotic Cell Features

Feature	Function
Nucleus	Area where most of the genetic material is stored and organized.
Nuclear envelope	Double membrane that encloses nucleus
Lysosome	Site where macromolecules are degraded
Ribosome	Site of polypeptide synthesis
Plasma membrane	Membrane that controls the movement of substances into and out of the cell, also the site of cell signaling
Cytosol	Site of many metabolic pathways and encloses all internal organelles.
Golgi apparatus	Site of modification, sorting and secretion of lipids and proteins
Peroxisome	Site where harmful toxins are broken down
Cytoskeleton	Protein filaments that provide shape and aid in movement
Mitochondrion	The site of ATP synthesis
Smooth ER	Site of detoxification and lipid synthesis
Rough ER	Site of protein sorting and secretion
Centrosome	Site where microtubules grow and centrosomes are found
Nuclear pore	Passageway for molecules to come into and go out of the nucleus
Nucleolus	Site for ribosome subunit assembly

Chromatin	A complex of protein and DNA
Chloroplast	Found in plant cells. Site of photosynthesis
Cell wall	Found in plant cells. Structure that provides cell support.
Central vacuole	Found in plant cells. Site that provides storage and regulates the cell volume.

The cytosol is the region of a eukaryotic cell that contains the membrane-bound organelles of the cell. Many different molecules are broken down in the cytosol. The sum of the chemical reactions by which cells utilize energy to sustain life and the breakdown and production of materials is called metabolism. Metabolic pathways are the series of steps that occur during metabolism. Enzymes are the proteins that are responsible for catalyzing, or accelerating these processes. There two types of metabolic processes. Catabolism is the breakdown of molecule into smaller molecules, and anabolism is the synthesis of molecules and macromolecules.

The cytoskelaton provides shape to the cell by the

use of microtubules, intermediate filaments, and actin filaments. Microtubles are hollow, cylindrical structures that are about twenty-five nanometers in diameter. They are polar structures that can grow and shorten in length during different phases of growth and retraction. These changes in length occur through a phenomenon called dynamic instability. In animal cells, microtubules extend out from the centrosome, and they extend out from various different points throughout the cell in plant cells and protists. The microtubules provide both organization to the organelles and cell shape. The intermediate filaments help to maintain cell shape and rigidity. These filaments tend to be relatively permanent as opposed to to the microtubules and actin fibers. The intermediate filaments are about ten nanometers in diameter. They also anchor cell and nuclear membranes. The actin filaments have the smallest diameter of the cytoskelatal filaments with a diameter of about seven nanometers. The actin filaments are spiral filaments that are composed of two strands of intertwined protein actin. The actin filaments provide cell shape and strength. They also play a role in muscle contractions and movement of the cell and cell cargo throughout the cell.

Motor proteins are a type of protein that interacts with cytoskeletal fibers through the use of ATP to produce movements of the cell and cell cargo along cytoskeletal filaments. Motor proteins pivot along a hinge to move molecules along fibers, contract fibers to produce cell movements, and change the shape of cell structures. Specialized motor proteins exist in the flagella and pila of cells to produce the whip-like action that produces movement of the cell. These motor proteins are anchored to filaments called basal bodies in the flagella and cilia of cells.

The nucleus of a cell is part of the cell that contains

most of the genetic information for reproduction of the cell. The nucleus is inclosed by a network of membranes called the endomembrane system that encloses the nucleus along with the endoplasmic reticulum, golgi apparatus, lysososomes, and peroxisomes. Materials are passed between the organelles of the endomembrane system by vesicles. Vesicles are small membrane enclosed spheres that carry materials.

The nuclear envelope is a double membrane structure that encloses the nucleus. Nuclear pores serve as openings that allow materials to enter and exist the nuclear envelope. Inside the nucleus, genetic material is stored on chromosomes that are compacted and stored inside the nucleus. Each chromosome is stored in non-overlapping chromosome territories which can be clearly seen when the cell is exposed to specific radioactive dyes. Additionally, the nucleolus serves the region of the nucleus in which ribosomes subunits are assembled.

The endoplasmic reticulum, also known as the ER, is a network of membranes that form flattened, fluid-filled tubules responsible for several functions. The fluid-filled tubules are called cisternae. The internal space inside the ER membranes is called lumen. There are two types of ER that are continuous. They are the rough ER and the smooth ER. Rough ER is studded with ribosomes on its surface while smooth ER lacks the ribosomes. The rough ER plays a role in protein sorting and transportation. Additionally, the rough ER attaches lipids and carbohydrates to proteins in a process called glycosylation. The smooth ER serves to remove many harmful toxins from the cell through metabolic reactions and the use of different enzymes.

The Golgi apparatus is a stack of flattened membranes that enclose a single compartment. The Golgi apparatus is responsible for the processing, sorting, and secretion of proteins along with continued glycosylation that had began in the endoplasmic reticulum. In the Golgi apparatus, enzymes called protease cut polypeptides in a process called proteolysis to create functional hormones. The Golgi apparatus also also packages cellular material into vesicles that follow secretory pathways to be released outside of the cell through the plasma membrane.

Lysosomes are another type of organelle that is responsible for the break down of various macromolecules in a cell. Lysosomes utilize acid hydrolases, which are hydrolytic enzymes that use a molecule of water to break covalent bonds. Lysosomes can breakdown various carbohydrates, proteins, lipids, and amino acids through acid hydrolases.

Vacuoles are prominent features in many plant cells, fungi, and some protists. Though vacuoles serve various functions, they are primarily responsible for storage of water, enzymes and inorganic ions. Most notably, a plant can be seen to wilt when the water in the central vacuoles of its cells is diminished. The central vacuole of a plant cell composes about 80% of the cell.

Peroxisomes are small organelles carry out various chemical processes in a cell including the break down of organic molecules. Peroxisomes are found in large numbers in the liver cells of an animal. The liver is responsible for the breakdown of numerous toxins that enter the body. Liver cells tend to get larger from the production of more peroxisomes in individuals who tend to drink large quantities of alcohol as the liver cells need to break down more toxins in the system. Peroxisomes utilize an enzyme called catalase to break down dangerous molecules such as hydrogen peroxide that can do a great deal of damage to a cell if left unchecked.

In addition to providing a cell with a state of

homeostasis, the plasma membrane of a cell serves several other functions. The plasma membrane attaches proteins to the phospholipid bi-layer to transport essential nutrients and ions into and out of the cell. The proteins are also used to signal other cells and allow the cell to adapt to changes in its environment. Lastly, the proteins attached to the plasma membrane of animal cells produce protein-protein interactions to promote cell adhesion that is critical in multi-cellular organisms for the cells to be able to work together to perform specific functions on a larger level.

Of the organelles in a cell, mitochondria and chloroplasts operate in a semi-autonomous manner. These organelles grow and divide in a manner similar to cells and contain genetic material. They do however depend on the other parts of the cell for some of their functions. Mitochondria look like a granule when examined through a light microscope and are similar in size to many bacteria.

The mitochondria can be found in both aniimal and plant cells. They serve to produce energy in the form of ATP for the cell. The mitochondria has both an inner and outer membrane. The region between the membranes is simply referred to as the intermembrane region. The inner membrane contains numerous folds that project into the inside of the mitochondria forming projections called cristae. These folds increase the surface area of the inner membrane allowing for more ATP to be produced. The cristae serve as the production sites for the ATP. The area that is inside the inner membrane is called the mitochondrial matrix. In addition to the production of ATP, mitochondria also play role in the breakdown of various other cellular molecules including hormones and fat cells that serve to provide heat to the cell.

Chloroplasts are organelles found in plant cells that use to energy obtain from light to synthesize molecules such as glucose in a process known as photosynthesis. Chloroplasts also have an outer and inner membrane with an intermembrane region between the membranes. Inside the two membranes, a third system of membranes esists called the thylakoid membrane that is composed of fluid-filled tubules. The tubules are stacked on top of each other to create a sturcture called a granum. Between the thylakoid membrane and the inner membrane, the compartment called the stroma exists that houses the granum.

Mitochondria and chloroplasts are unique in that they have their own DNA that is separate from the DNA found in the nucleus. This DNA is much like the DNA of bacteria in that it is circular in shape as opposed to the linear DNA found in the nucleus. This DNA is similar to DNA of bacteria which has led to some interesting theories on how eukaryotic cells evolved to have mitochondria and chloroplasts. The endosymbiosis theory asserts that chloroplasts and mitochondria were formed when eukaryotic cells engulfed bacteria. Instead of breakdown and absorbing the bacteria, the bacteria and eukaryotic cells formed a relationship where the bacteria remained in the cell in a state where both benefited from the presence of the other. This type of interaction where one organism lives inside a larger organism is called endosymbiosis.

When considering the various parts of the cell and how they interact, one can clearly see that there is a great deal of complication to the design of a cell. The interactions a small level can clearly be noted as affecting successively larger structures in the organization of life ranging from the cellular level to the biosphere. It can also be seen that the interaction is dynamic in that as one system affects change, both the larger system and smaller systems of the hierarchical structure of life can be seen as being affected. Through out this book, a basis has been given for understanding the fundamental principles of biology and life along with how different levels of life build on each other.

www.ingramcontent.com/pod-product-compliance
Lightning Source LLC
Chambersburg PA
CBHW020956180526
45163CB00006B/2393